Food Chains in a DESERT HABITAT

ISAAC NADEAU
Photographs by
DWIGHT KUHN

The Rosen Publishing Group's
PowerKids Press™
New York

To E.G., Luc, and Brady, oases all — Isaac Nadeau
To Heather — Dwight Kuhn

Published in 2002 by The Rosen Publishing Group, Inc.
29 East 21st Street, New York, NY 10010

First Edition

Book Design: Emily Muschinske
Project Editor: Emily Raabe

Photo Credits: pp. 4, 5 (background and cactus), 7, 8, 9, 10, 11 (large picture), 13, 17, 19, 22 © Tomas Weiwandt; all other photographs © Dwight Kuhn.

Nadeau, Isaac.
Food chains in a desert habitat / Isaac Nadeau.
p. cm. — (The library of food chains and food webs)
Includes bibliographical references (p.).
ISBN 0-8239-5760-8
1. Desert ecology—Juvenile literature. 2. Food chains (Ecology)—Juvenile literature. [1. Desert ecology. 2. Food chains (Ecology) 3. Deserts. 4. Ecology.] I. Title. II. Series.
QH541.5.D4 N24 2002
577.54'16—dc21

2001000278

Manufactured in the United States of America

Contents

Food Chains in the Desert

Every living thing in the desert is part of a food chain. A food chain is a way of showing how energy is passed on when one living thing eats another.

A roadrunner chases a lizard. The lizard races beneath a cactus and dives to safety in a crack in a rock. Why was the roadrunner chasing the lizard? How did the lizard get away? The answer to both questions is the same. It is energy. It takes energy for the roadrunner to chase the lizard, and the lizard uses energy to run away. All animals get their energy from what they eat. A roadrunner gets its energy by eating lizards and other animals. A lizard gets its energy from eating ants and other insects. Ants get their energy from eating animals and plants. Plants get their energy from sunlight. The plants, ants, lizard, and roadrunner all are parts of a desert food chain.

In this desert food chain, you can follow the energy as it moves from cactus to grasshopper, grasshopper to grasshopper mouse, and grasshopper mouse to rattlesnake.

Habitat, Sweet Habitat

A habitat is the place where a plant or an animal lives. When you imagine the desert habitat, what do you picture? Do you think of rocks and sand? Do you imagine dust and thorns? Do you just imagine the hot sun beating down? All of these things are part of the desert habitat, but if you look more closely, you will find that a desert is not as deserted as it appears. In fact, it is a home for all kinds of amazing plants and animals. Mesquite trees grow deep roots near streambeds, called **arroyos**. A cactus wren makes its nest in the arms of a **saguaro** cactus. A scorpion carries her babies on her back. These creatures have found ways to live in a dry land. They find everything they need in the desert habitat, including food, shelter, and even water.

These tall saguaro cacti store water in their stems. They also save water by not having any leaves. Plants usually lose water through their leaves.

Where Is the Water?

In the desert habitat, water is one of the greatest treasures. Everything needs water, whether it is a pupfish swimming in a desert spring, or a turkey vulture flying high over the land. Deserts only get a small amount of rain each year. When it does rain, the hot sun causes much of the water to go quickly back into the air. This is called **evaporation**. With so little water to go around, plants and animals need ways to find water and keep it. Desert plants and animals have many tricks to get water. The **ocotillo** plant, for example, has cup-shaped leaves. When it rains, the ocotillo collects the rain in its leaves. The kangaroo rat would get very thirsty if it came out during the hot desert day. Instead it moves about only at night, when the air is cool. The kangaroo rat is so good at using water that it never even needs to take a drink! All the water it needs comes from the seeds it eats.

Sometimes it does rain in the desert, which allows creatures such as this ocotillo plant (bottom left) *and this kangaroo rat* (top right) *to survive in this harsh environment.*

Please Pass the Sun!

If the sun was your favorite food, the desert would be a great place to live. Sometimes, though, there can be too much of a good thing. There is more sunlight in the desert than plants can use. Some desert plants, like the ocotillo, lose their leaves when there is too much sun and grow them back when the air is cooler.

Plants are the first link in any desert food chain. They take the energy from sunlight and make it into food. For this reason, plants are called **producers**. Plants produce sugar by mixing air, water, and sunlight in their leaves and stems. This is called **photosynthesis**. Plants use sugar to grow their roots, stems, and flowers. When an animal, such as a jackrabbit, eats part of a plant, it gets some of the energy that the plant produced.

Most desert plants have **adapted** to living in the sun. Some plants grow in the shade of rocks or other plants. Many desert plants are covered with wax, to keep them from losing water through evaporation.

These flowering desert plants have adapted to living in the bright sunlight of the desert.

A Desert Is Not Deserted

At first, the desert might look like a land without animals. If you walk quietly and look closely, you will find that a desert is full of animals. Almost anywhere you see plants in the desert, you will find signs of animals, too. That is because there are many animals in the desert that depend on plants for their food. Animals that eat only plants for food are called **herbivores**. Herbivores are the second link in a desert food chain. They use the energy from the plants to build their homes, look for food, reproduce, and run from **predators**. Jackrabbits, grasshoppers, mule deer, and hummingbirds are just a few of the herbivores that live in desert habitats.

Jackrabbits lose heat through their large ears. This helps them to stay cool in the heat of the desert.

Hunting in the Desert

Poisonous fangs and stingers, sharp claws and teeth, and speedy legs and silent wings are all tools of a predator. There are many predators in the desert habitat. Animals that eat other animals are called **carnivores**. Carnivores are the third link in a desert food chain. A rattlesnake eating a kangaroo rat is a carnivore. Desert carnivores include birds, snakes, lizards, spiders, insects, and **mammals**. Roadrunners are very fast runners and fearless hunters. They chase their **prey** on foot and catch it in their beaks. Roadrunners even catch rattlesnakes and scorpions. Most desert predators are also prey to other carnivores. A carnivore that eats another carnivore is called a secondary carnivore. Secondary carnivores are the fourth link in a desert food chain.

This banded gecko hides from the heat during the day, coming out at night to hunt insects.

Although this tarantula (right) and this scorpion (below) are fearsome carnivores in the desert, many other desert carnivores, such as owls, bats, and roadrunners, hunt them and eat them.

Nothing Wasted

Sometimes an animal dies and is not eaten right away, or part of the animal is eaten and the rest is left behind. In the desert, dead animals do not go very long without being noticed by a hungry **scavenger**. Scavengers are animals that eat the bodies of dead animals. Coyotes and turkey vultures are both scavengers. Coyotes are scavengers that are famous for being able to eat almost anything. Animals that eat both plants and animals are called **omnivores**. Scavengers such as the turkey vulture and the coyote play an important role in desert food chains because they keep the energy, **nutrients**, and water in dead animals from going to waste.

Coyotes have been known to eat deer, sheep, jackrabbits, mice, grasshoppers, dead birds, mammals, berries, and hundreds of other kinds of food.

Scavengers, such as this long-horned cactus beetle that is eating a grasshopper mouse (above), and this coyote (below), help to keep the desert habitat clean by eating dead animals.

Desert Decay

(Above) ***In life, this scorpion was a dangerous predator. In death, it will feed producers as a part of the desert soil.***

Why do plants have roots? They use their roots to get water from the soil. Plants also use their roots to get nutrients. Nutrients are mixed with the water in the desert soil. The water comes from the rain, but from where do the nutrients come? Nutrients become part of the desert soil because of the hard work of a group of living things called **decomposers**. Decomposers in the soil include **bacteria**, and fungi such as mushrooms. Fungi help plants by causing small bits of dead plants and animals to rot. Some special fungi grow right on the roots of desert plants. These fungi help the plants get nutrients, and the plants help give the fungi energy.

In most desert soils, decay occurs very slowly. This is because decay usually happens best where there is water. With the help of the decomposers, however, every desert food chain will, in time, end with nutrients being returned to the desert soil.

As this dead saguaro cactus decays, the nutrients in its body will go back to the soil.

A Worldwide Web of Life

What would happen if it didn't rain in the desert? The plants would have no water and would die. What would happen if the plants, whose seeds mice eat, all died? The mice would starve. What would happen if all the mice in the desert disappeared? The coyotes, owls, and hawks would go hungry. The plants and animals in a desert habitat depend on each other. The mouse depends on the seeds it gets from the plants. The owl depends on the mouse, and so on. The food chains in a desert habitat all are connected in a web of life. If something happens to one part of the web, it will have an effect on the other parts.

Color Key

- carnivores
- decomposers
- herbivores
- omnivores
- producers

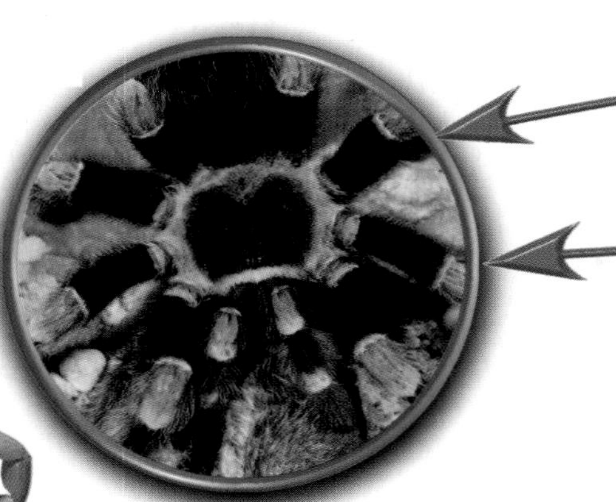

Exploring the Desert

For thousands of years, people have been exploring and learning new things about the desert, but there is always more to be learned about this amazing habitat.

If you ever have a chance to visit a desert habitat, you might not want to leave (unless you forget your water bottle!). There is more to explore in the desert than you could imagine. One day you might discover the treasures stored in a pack rat's home. Some of these are hundreds of years old! Another day you might be caught in a desert rainstorm, with water rushing down normally dry arroyos. In the spring, you might see the desert come alive with wildflowers. At night you could see a bat drinking from the white flower at the top of a saguaro cactus. There are always wonderful things going on in the desert.

Glossary

adapted (uh-DAPT-ed) Changed to fit conditions.
arroyos (uh-ROY-ohz) Streambeds carved by water.
bacteria (bak-TEER-ee-uh) Tiny living things that are seen with a microscope.
carnivores (KAR-nih-vorz) Animals that eat other animals for food.
decomposers (dee-kum-POH-zerz) Organisms that break down the bodies of dead animals and plants.
evaporation (ee-va-puh-RAY-shun) Water lost to the warm, dry air.
herbivores (ER-bih-vorz) Animals that eat plants and plant matter.
mammals (MAM-mulz) Warm-blooded animals that have a backbone, are often covered with hair, breathe air, and feed milk to their young.
nutrients (NOO-tree-ints) Anything that a living thing needs for its body to live and grow.
ocotillo (oh-koh-TEE-yoh) A desert plant with tall, thin stems that loses its leaves in dry times.
omnivores (AHM-nih-vorz) Animals that eat both plants and animals.
photosynthesis (foh-toh-SIN-thuh-sis) The process in which plants use energy from sunlight, gases from air, and water to make food.
predators (PREH-duh-terz) Animals that kill other animals for food.
prey (PRAY) An animal that is hunted by another animal for food.
producers (pruh-DOO-serz) Plants, which use sunlight to make energy available for use by living things.
saguaro (sah-WAHR-oh) A type of large cactus.
scavenger (SKA-ven-jur) An animal that feeds on dead animals.

Index

Web Sites

To learn more about food chains in a desert habitat, check out this Web site:
www.ran.org/ran/kids_action